Astronomy Now!™

A Look at
JUPITER

Suzanne Slade

PowerKiDS
press.
New York

With love, to Fred and Diane Buckingham

Published in 2008 by The Rosen Publishing Group, Inc.
29 East 21st Street, New York, NY 10010

First Edition

Editor: Amelie von Zumbusch
Book Design: Greg Tucker
Photo Researcher: Nicole Pristash

Photo Credits: Cover, p. 13 (main) © SuperStock, Inc.; pp. 5 (main), 7 (inset), p. 17 (main) by PhotoDisc; pp. 5 (inset), 13 (top inset, bottom inset), 15, 19 (main), 21 © Getty Images; p. 7 (main) © Shutterstock.com; p. 9 Courtesy NASA/JPL/Space Science Institute; p. 11 © Mark Garlick/Photo Researchers, Inc.; p. 12 © AFP/Getty Images; pp. 17 (inset), 19 (inset) Courtesy NASA/JPL-Caltech.

Library of Congress Cataloging-in-Publication Data

Slade, Suzanne.
  A look at Jupiter / Suzanne Slade. — 1st ed.
      p. cm. — (Astronomy now!)
  Includes index.
  ISBN-13: 978-1-4042-3829-9 (lib. bdg.)
  ISBN-10: 1-4042-3829-8 (lib. bdg.)
  1. Jupiter (Planet)—Juvenile literature. 2. Jupiter (Planet)—Atmosphere—Juvenile literature. I. Title.
  QB661.S63 2008
  523.45—dc22
                                            2007007137

Manufactured in the United States of America

Contents

The Largest Planet

Jupiter is the biggest planet in our **solar system**. More than 1,300 Earths could fit inside this huge planet. On a clear night, you can search for Jupiter in the sky. Jupiter looks like a bright star. However, Jupiter does not burn, as stars do. The Sun's light makes Jupiter shine. Jupiter is the second-brightest planet, after Venus.

People have noticed the bright light of Jupiter moving across the night sky for thousands of years. Over time, Jupiter has been known by many different names. Today, we call the planet Jupiter after Jupiter the king of the Roman gods.

Jupiter is a huge planet. It weighs more than all the other planets in our solar system put together. *Inset:* The god Jupiter, after whom the planet is named, is also known as Jove.

Jupiter on the Move

Jupiter is the fifth planet from the Sun. The four planets closest to the Sun are known as inner planets. Jupiter is the first of the four outer planets. Jupiter moves in an **oval** path around the Sun, called an orbit. It takes about 12 Earth years for Jupiter to travel once around the Sun.

Jupiter also spins around a pretend line through its center, called an axis. Jupiter spins faster than any other planet. Jupiter makes one full turn in only 9 hours and 50 minutes. This length of time is one day on Jupiter.

Jupiter's orbit lies between the orbits of the planets Mars and Saturn. *Inset:* Jupiter is more than 700 million miles (1.1 billion km) from the Sun.

A Gas Giant

If you traveled to Jupiter, you would not need landing gear. This is because Jupiter is a gas planet. Gases are not hard enough to land on. Jupiter is mostly made of gases called **hydrogen** and **helium**. It has water in the form of gas, too.

Gases also form Jupiter's **atmosphere**, a 600-mile- (966 km) deep cover of gases that sits on the planet. Fast-moving clouds in Jupiter's atmosphere make beautiful orange and white stripes. Warm gases traveling from inside Jupiter make white stripes, called zones. Cold gases moving toward Jupiter's center cause the belts, or darker stripes.

All the colors and shapes you can see in this picture of Jupiter are clouds in the planet's atmosphere. While Earth's clouds are all made from water, Jupiter has clouds made of both water and other kinds of matter.

9

Inside Jupiter

Jupiter's core, or center, is mostly made of different metals. It also has matter called **silicon**. No one is sure exactly what Jupiter's core is like. Some **astronomers** think that it is hard and rocky. Others believe the core is melted and is like a hot, thick soup.

Jupiter's core is about 7,500 miles (12,000 km) wide. This is about the same size as the planet Venus. Astronomers believe that Jupiter's huge core is between 36,000 and 50,000° F (20,000-28,000° C). This makes it even hotter than the outside of the Sun!

Jupiter's core is gray and white in this drawing. The light blue around the core is liquid metallic, or metal-like, hydrogen. The darker blue beyond that is liquid hydrogen that is not metallic.

Cool Facts

Jupiter is so huge that all the other planets in our solar system could fit inside it.

Most planets are round but not Jupiter! It spins so fast that the middle of the planet becomes fat, and the top and bottom of the planet flatten out.

Not all Jupiter's moons have names. Newly discovered moons are given a number. Later, scientists give a name to each moon.

The spacecraft *Pioneer 10* visited Jupiter. It carried messages and pictures of people and of Earth, in case it is found by beings from another planet.

Pioneer 10

A Jupiter Timeline

1995 – The spacecraft *Galileo* reaches Jupiter. *Galileo* studies the planet's moons and atmosphere.

1979 – *Voyager 1* visits Jupiter. It studies the planet and discovers Jupiter's rings.

Fun Figures

Jupiter is 88,846 miles (142,984 km) across.

One year on Jupiter lasts 11 years and 313 Earth days. This is how long it takes Jupiter to orbit the Sun once.

As Jupiter and Earth orbit the Sun, the space between them changes. At their closest point, Jupiter and Earth are 366 million miles (589 million km) from each other.

1973 – The spacecraft *Pioneer 10* flies by Jupiter and takes the first close-up pictures of the planet.

1610 – Galileo Galilei discovers Jupiter's four biggest moons.

13

Stormy Weather

Jupiter's atmosphere is full of stormy and windy weather. Sometimes lightning flashes across its clouds. It has strong winds that can blow up to 400 miles per hour (644 km/h). Spinning storms are also found in Jupiter's atmosphere.

The largest storm on Jupiter is called the Great Red Spot. An astronomer named Robert Hooke discovered the Great Red Spot in 1664. The twisting winds of this storm continue to blow today. The size of the Great Red Spot often changes. Sometimes it is the size of Earth. Other days it grows as large as three Earths.

The big red oval in this picture is Jupiter's Great Red Spot. The smaller white oval beneath it is also a big, spinning storm.

All Around Jupiter

Jupiter has three rings that circle around it. These rings are mostly made of space dust. Moons also orbit Jupiter. Jupiter has more moons than any other planet. Astronomers have found 63 moons so far, and new ones are discovered all the time. Jupiter's smallest moon is about 1 mile (1.6 km) wide. The largest is more than 3,000 miles (4,828 km) wide. This means it is bigger than the planet Mercury.

In 1610, Galileo Galilei discovered Jupiter's four largest moons. He made a tool with powerful lenses called a telescope to see them. These four moons are called the Galilean moons.

Many people think Jupiter and its moons look like a mini solar system. *Inset:* The yellow lines on the left of this picture are Jupiter's rings. The blue line is the edge of Jupiter.

17

Io and Europa

The Galilean moon closest to Jupiter is called Io. Io is a little larger than Earth's Moon. There are at least nine live **volcanoes** on Io. Every year, thousands of tons (t) of **lava** stream from these volcanoes. The shape of Io changes as new lava continues to flow.

The next Galilean moon, Europa, is 417,000 miles (671,000 km) from Jupiter. Europa is covered with ice. This white and gray moon looks smooth and has many thin dark lines. Astronomers believe these lines were formed when water from inside Europa pushed through the ice that covers this moon.

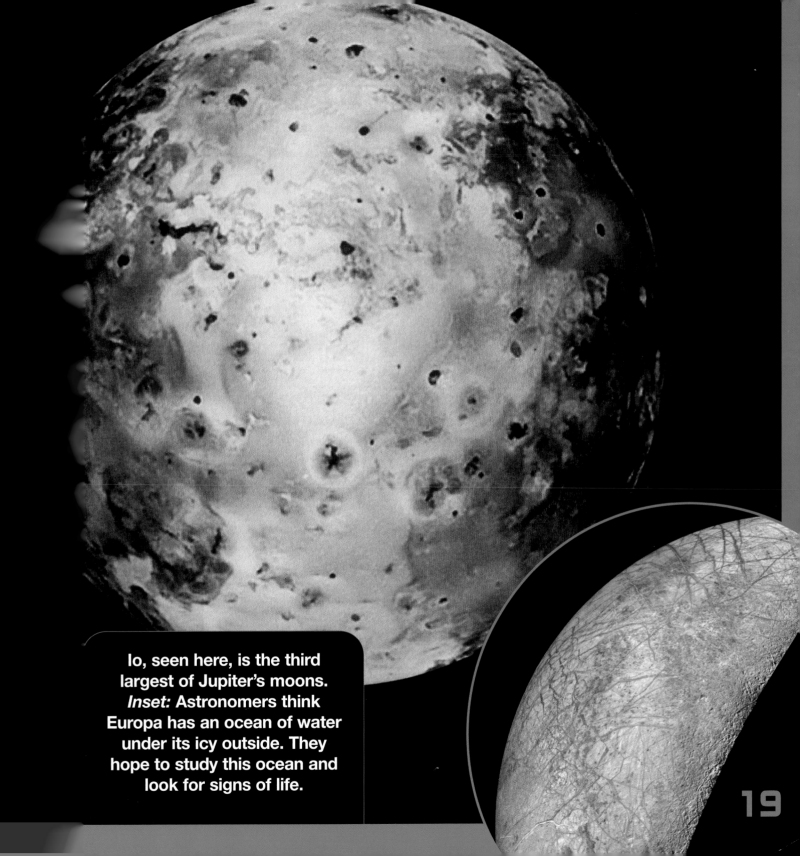

Io, seen here, is the third largest of Jupiter's moons. *Inset:* Astronomers think Europa has an ocean of water under its icy outside. They hope to study this ocean and look for signs of life.

Ganymede and Callisto

Jupiter's third Galilean moon is called Ganymede. It is the largest moon in the solar system. Ganymede is 3,280 miles (5,279 km) wide. Astronomers think the outside of Ganymede is covered with rocks and ice. Ganymede has large, deep holes, called craters. Craters form when big rocks crash into a planet or a moon.

Callisto is the Galilean moon farthest from Jupiter. It is made of rock and ice. There are more craters on Callisto than on any other object in the solar system. Callisto is the darkest Galilean moon, but it is still two times brighter than Earth's Moon.

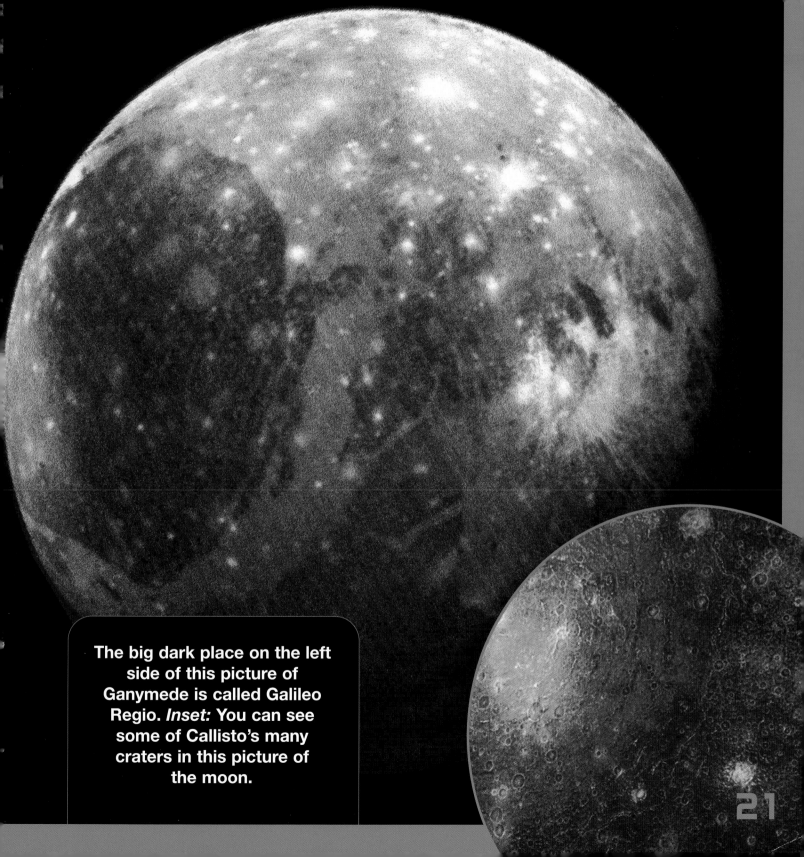

The big dark place on the left side of this picture of Ganymede is called Galileo Regio. *Inset:* You can see some of Callisto's many craters in this picture of the moon.

Traveling to Jupiter

Astronomers today learn about Jupiter from spaceflights. The first spacecraft to fly past Jupiter was *Pioneer 10*. It left Earth March 2, 1972, and reached Jupiter 21 months later. *Pioneer 10* took close-up pictures of Jupiter. In the following years, other spacecraft traveled to Jupiter and studied its storms, wind, and lightning.

A spacecraft, named *Juno*, will leave for Jupiter in 2011. It will study Jupiter's atmosphere and core. In 2015, another spacecraft, called *Europa Geophysical Explorer*, will search for places to land on Europa. Someday, people may study Jupiter while living on one of its moons!

Glossary

astronomers (uh-STRAH-nuh-merz) People who study the Sun, the Moon, the planets, and the stars.

atmosphere (AT-muh-sfeer) The gases around an object in space.

helium (HEE-lee-um) A light, colorless gas.

hydrogen (HY-dreh-jen) A colorless gas that burns easily and weighs less than any other known kind of matter.

lava (LAH-vuh) Hot, melted rock that comes out of a volcano.

oval (OH-vul) A shape that looks like a circle with two sides pressed in.

silicon (SIH-lih-kun) A kind of matter found in rocks and sand.

solar system (SOH-ler SIS-tem) A group of planets that circles a star.

volcanoes (vol-KAY-nohz) Openings in a planet or moon that sometimes shoot up hot, melted rock called lava.

Index

Web Sites

Due to the changing nature of Internet links, PowerKids Press has developed an online list of Web sites related to the subject of this book. This site is updated regularly. Please use this link to access the list:
www.powerkidslinks.com/astro/jupiter/